1 MONTH OF
FREE
READING

at

www.ForgottenBooks.com

By purchasing this book you are eligible for one month membership to ForgottenBooks.com, giving you unlimited access to our entire collection of over 1,000,000 titles via our web site and mobile apps.

To claim your free month visit:

www.forgottenbooks.com/free896691

ISBN 978-0-265-83618-7
PIBN 10896691

This book is a reproduction of an important historical work. Forgotten Books uses
state-of-the-art technology to digitally reconstruct the work, preserving the original format
whilst repairing imperfections present in the aged copy. In rare cases, an imperfection in
the original, such as a blemish or missing page, may be replicated in our edition. We do,
however, repair the vast majority of imperfections successfully; any imperfections that
remain are intentionally left to preserve the state of such historical works.

BULLETIN OF THE
U.S.DEPARTMENT OF AGRICULTURE
No. 48

Contribution from the Bureau of Plant Industry, Wm. A. Taylor, Chief.
December 19, 1913.

THE SHRINKAGE OF SHELLED CORN WHILE IN CARS IN TRANSIT.

By J. W. T. DUVEL, *Crop Technologist in Charge of Grain Standardization Investigations,* and LAUREL DUVAL, *formerly in Charge of the Grain Standardization Laboratory at Baltimore, Md.*

INTRODUCTION.

In January, 1910, special investigations were begun at Baltimore, Md., to determine the amount of natural shrinkage or loss in weight of shelled corn containing various percentages of moisture during transit in cars and while in storage in elevators.[1]

The data on shrinkage contained in these pages comprise the results of four special shipments of corn from Baltimore, Md., to Chicago, Ill., and return. The first experiment, made April 15, 1910, was in cooperation with the Baltimore & Ohio Railroad Co. The second, third, and fourth experiments were made in cooperation with the Pennsylvania Railroad Co. The second shipment was made on December 24, 1910; the third, March 2, 1911; and the fourth, May 11, 1911. The losses in weight shown for the various shipments do not include the losses in loading or unloading, but simply the natural shrinkage in the weight of the corn while in the cars.

METHODS USED IN DETERMINING SHRINKAGE IN CORN.

The freight cars used in these experiments were especially selected, uniform in design and condition, and specially coopered to make them grain tight. The cars were held together on the track for several days prior to loading, so that there would be no variation in their weight due to differences in the condition of the wood. An empty box car of the same series and condition was moved in the same train with the loaded cars, as a check in determining the variation in weight due to the absorption or evaporation of moisture by the cars

---

[1] The results of the first experiment of the series on the shrinkage of shelled corn in storage were published as Circular 81, Bureau of Plant Industry, U. S. Department of Agriculture.

under varying weather conditions. A standard scale test car was also attached to each shipment, and all scales were carefully tested prior to weighing the cars. Each shipment was accompanied by a representative of the United States Department of Agriculture, for the purpose of making weighings en route at certain division points and of keeping a complete record of the changes in the temperature of the corn. The temperature records were taken by means of electrical resistance thermometers, which were placed in the corn at the time of loading, as shown in figure 1. In all of the shipments, with the exception of the first, two cars were loaded from each lot of corn. In the second, third, and fourth shipments one car from each lot of

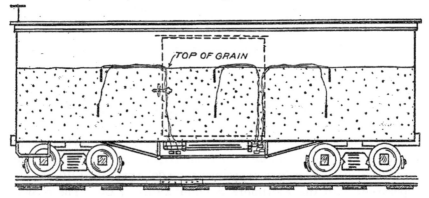

FIG. 1.—Sectional view through the center of a freight car, showing the position of the six electrical resistance thermometers in the stored corn.

corn was shipped from Baltimore to Chicago and return, while the duplicate cars of corn from each lot, together with an empty box car of the same type, were held on the track in the yards at Baltimore.

FIRST EXPERIMENT, APRIL 14 TO MAY 11, 1910.

CORN LOADED INTO CARS.

On April 14, 1910, five cars were loaded with shelled corn from the Locust Point elevators of the Baltimore & Ohio Railroad Co., Baltimore, Md. The amount of corn contained in each car varied from 65,920 to 67,160 pounds, with an average of 66,832 pounds, or slightly more than 1,193 bushels. The different lots of corn used in this experiment were taken from the regular stock in the elevators and were selected primarily with the view to having corn with a relatively wide range in moisture content. Before loading, each lot was thoroughly mixed by handling in the elevator, so that the quality and condition of the corn would be uniform throughout the car. At the time of loading, each car was equipped with six electrical resistance thermometers, as shown in figure 1. These thermometers were placed in the corn in different parts of the car, in order that the temperature of the corn at those points could be ascertained at any time during the experiment without opening the car, the thermometers being connected with wire leads which extended to the outside of the car. The average condition and quality of the corn in each car at the beginning of the experiment, as represented by the analyses of samples taken when the corn was loaded into the cars, is shown in Table I.

TABLE I.—*Condition and quality of the corn when loaded into each car on April 14, 1910, for the first experiment.*

Car No.	Moisture content.	Sound corn.	Weight per bushel.	Cob, dirt, etc.	Badly broken kernels.
	Per cent.	*Per cent.*	*Pounds.*	*Per cent.*	*Per cent.*
1	19.8	93.2	53.1	1.1	5.0
2	18.6	95.9	53.6	.3	3.1
3	17.8	96.0	54.1	.7	4.9
4	17.4	97.4	55.0	.7	4.2
5	16.7	94.1	54.0	1.3	5.7

As will be seen by Table I, the average moisture content of the corn when loaded into car No. 1 was 19.8 per cent. The corn in car No. 2 contained an average of 18.6 per cent of moisture. The corn in car No. 3 was a mixture of the same kind of corn as that contained in cars No. 2 and No. 5, the average moisture of the mixture being 17.8 per cent. The average moisture content of the corn in car No. 4 was 17.4 per cent, and the corn in car No. 5, which had been artificially dried, contained an average of 16.7 per cent of moisture.

SHRINKAGE IN THE WEIGHT OF THE CORN.

The amount of corn placed in each car and the shrinkage in weight while the corn was in the cars from April 14 to May 11, 1910, together with the average temperature of the corn in each car at the time of loading and unloading, are shown in Table II.

TABLE II.—*Shrinkage, or loss in weight, and temperature changes on each of 5 cars of corn used in the first experiment.*

Car No.	Corn when loaded.		Shrinkage, or loss in weight.		Average temperature of corn when—	
	Moisture content.	Net weight.	Pounds.	Per cent.	Loaded.	Unloaded.
	Per cent.	*Pounds.*			*° F.*	*° F.*
1	19.8	67,130	2,450	3.65	58	139½
2	18.6	67,120	320	.48	52	84
3	17.8	65,920	290	.44	54	82½
4	17.4	67,160	180	.27	54	58½
5	16.7	66,940	120	.18	58	62

It will be seen by Table II that the loss in weight of the corn from the time the cars were loaded on April 14 until they were unloaded May 11, 1910, varied according to the moisture content. The corn in car No. 1, which contained an average moisture content of 19.8 per cent, showed a total shrinkage in weight of 2,450 pounds, or 3.65 per cent. The natural shrinkage in the weight of the corn in car No. 3, which contained an average moisture content of 17.8 per cent, showed a loss of 290 pounds, or 0.44 of 1 per cent. The corn in car No. 4, with 17.4 per cent of moisture, lost 0.27 of 1 per cent, while the corn in car No. 5, which was artificially dried corn and contained 16.7 per cent of moisture, showed a shrinkage of 120 pounds, or 0.18 of 1 per cent.

The heavy loss occurring in car No. 1 was due to the fact that this corn showed a marked deterioration during the experiment. It was hot, sour, and discolored at the time of unloading, the average temperature of the corn being 139.5° F. These same factors, with the addition of the records of the mean daily air temperature through which the cars passed while en route from Baltimore to Chicago and return, and also the average temperature of the corn during the period covered by the experiment, are graphically presented in figure 2. By examining the data shown in this diagram it will be seen that the natural shrinkage increased with the increase in the moisture content of the different lots of corn; likewise, there is a close correlation between the moisture content of the corn and the increase in temperature resulting from the deterioration of the high-moisture corn.

The average temperature of the corn in the five cars immediately after loading varied from 52° to 58° F., or practically the same as the air temperature when the corn was loaded. The corn in car No. 1 showed an increase in temperature of 15 degrees from April 15 to April 20, while the temperature of the corn in the other four cars during this time remained practically the same as when loaded. The increase in temperature in car No. 1 was undoubtedly caused by the fact that the corn in that car contained the highest percentage of moisture, 19.8 per cent. The relatively warm air temperature which prevailed at that time, together with the high moisture content of the corn, afforded favorable conditions for fermentation and the development of molds, resulting in a corresponding deterioration of the corn. On April 21 there was an increase of 6 degrees in the temperature of the corn in car No. 1 over the previous day. At this time the cars were passing through a much colder atmosphere, the mean daily air temperature decreasing from 52° on April 20 to 36° F. on April 21. During the same period the temperature of the corn in the other four cars did not show an increase; in fact, the temperature of the corn in two of the cars decreased slightly.

Figure 2 also shows that the mean daily air temperature from April 21 to April 27 was much lower than the average temperature of the corn in any of the cars, varying from 36 to 40 degrees. As a result of this low air temperature the temperature of the corn in all of the cars decreased slightly, except that in car No. 1, which

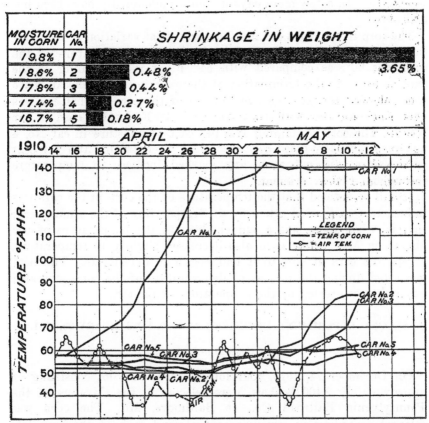

FIG. 2.—Diagram showing the shrinkage in weight of the corn in each of five cars in transit from Baltimore to Chicago and return, the average temperature of the corn in each car, and the mean daily air temperature through which the cars passed from April 14 to May 11, 1910.

showed an increase form 79° on April 21 to 135.5° F. on April 27, the deterioration being so rapid as to overcome or conceal any influence of the air temperature. The highest temperature of the corn in car No. 1 was 142° F., which was reached on May 3. On May 7 the average temperature was 139° F. This temperature prevailed until May 11, at which time the corn was unloaded and found to be very badly damaged.

The corn in car No. 2 did not show any appreciable increase in temperature until May 4, at which time the temperature of the corn was 61° F., an increase of 7 degrees over the record of the previous day. This increase in temperature was probably influenced to some extent by the air temperature, the mean daily air temperature having increased from 52° on May 2 to 62° F. on May 3. The temperature of the corn in this car at the end of the experiment on May 11 was 84° F. In a few more days it would undoubtedly have been hot and sour.

The corn in car No. 3 did not show any great variation in temperature until May 6, at which time the average temperature of the corn was 60°, as compared with 54° F. at the beginning of the experiment. At the termination of the experiment, when car No. 3 was unloaded the average temperature of the corn in that car was 82.5°, as compared with an average of 84° F. for the corn in car No. 2. The corn containing 17.4 per cent of moisture in car No. 4 and the artificially dried corn in car No. 5, containing 16.7 per cent of moisture, did not show any marked increase in temperature throughout this experiment.

It will be noticed that the temperature of the corn in the cars was influenced to a limited extent by the temperature of the atmosphere; that is, when there was a sudden drop in the temperature of the air, as on May 5, the corn also decreased slightly in temperature, except the cars of corn that had begun to deteriorate. The heat generated during the process of deterioration of the corn concealed any influence exerted by the temperature of the air.

SECOND EXPERIMENT, DECEMBER 24, 1910, TO JANUARY 20, 1911.

CORN LOADED INTO CARS.

On December 24, 1910, four lots of corn, each lot consisting of approximately 2,400 bushels, were selected at the elevators of the Pennsylvania Railroad Co., at Baltimore, Md. The corn was selected according to moisture content, and each of the four lots was first thoroughly mixed in the elevator before loading into cars. Two cars were loaded from each lot, one being forwarded from Baltimore to Chicago and return, over the Pennsylvania lines via the Fort Wayne route, while the duplicate car of each lot was held on the track in the Canton yards at Baltimore. Eight cars consequently were included in the experiment, in order to study the effect of atmospheric conditions on the shrinkage in weight and rate of deterioration. The cars held on the track at Baltimore were weighed, and temperature records were taken on the same days that weighings and temperature readings were made on the cars in transit. These eight cars were each equipped with six electrical resistance thermometers, the relative position of which is shown in figure 1.

The average condition and quality of the four lots of corn at the beginning of the experiment, as represented by the results of the analyses of samples taken when the corn was being loaded into cars, are shown in Table III.

TABLE III.—*Factors showing the condition and quality of each lot of corn in the second experiment when loaded into cars on December 24, 1910.*

Lot No.	Car designation and movement.	Moisture content.	Sound corn.	Weight per bushel.	Cob, dirt, etc.	Badly broken kernels.
		Per cent.	Per cent.	Pounds.	Per cent.	Per cent.
1.....	Car 1–C, transit, Baltimore to Chicago and return................................... Car 1–B, held on track in Baltimore yards....	22.0	94.3	50.9	0.2	2.5
2.....	Car 2–C, transit, Baltimore to Chicago and return................................... Car 2–B, held on track in Baltimore yards....	19.0	96.9	54.1	.1	3.5
3.....	Car 3–C, transit, Baltimore to Chicago and return................................... Car 3–B, held on track in Baltimore yards....	17.0	98.2	55.4	.3	3.1
4.....	Car 4–C, transit, Baltimore to Chicago and return................................... Car 4–B, held on track in Baltimore yards....	13.3	97.9	56.2	.4	12.1

It will be seen by reference to Table III that the average moisture content of the corn in lot No. 1 was 22 per cent; lot No. 2, 19 per cent; lot No. 3, 17 per cent; and lot No. 4, 13.3 per cent—a range of 8.7 per cent. Special attention is also called to the low weight per bushel of the corn from lot No. 1, containing 22 per cent of moisture, and the high percentage of the badly broken corn in lot No. 4. Much of this broken corn was of the consistency of coarse flour or meal.

SHRINKAGE IN THE WEIGHT OF THE CORN.

The quantity of corn placed in each car, the shrinkage in weight while in the cars from December 24, 1910, to January 20, 1911, together with the average temperature of the corn in each car at the time of loading and unloading, are shown in Table IV. These same data are presented diagrammatically in figure 3. In addition to the natural shrinkage in weight and the other data given in Table IV, figure 3 also shows the mean daily temperature of the air through which the cars passed while en route from Baltimore to Chicago and return, together with the average temperature of the corn in each car, as compared with like observations made on the corresponding lots of corn held on the track in the railroad yards at Baltimore.

TABLE IV.—*Weight of corn in each car immediately after loading on December 24, 1910, and the shrinkage, or loss in weight, together with the moisture content at time of loading and the average temperature of the corn at time of loading and unloading.*

Lot No.	Car designation and movement.	Moisture content of corn when loaded.	Net corn in car when loaded.	Shrinkage, or loss in weight.		Average temperature of corn when—	
				Pounds.	Per cent.	Loaded.	Unloaded.
		Per cent.	*Pounds.*			° *F.*	° *F.*
1.....	Car 1–C, transit, Baltimore to Chicago and return.....................	22.0	67,140	180	0.27	31.0	33.0
	Car 1–B, held on track in Baltimore yards.......................		67,140	230	.34	31.0	38.2
2.....	Car 2–C, transit, Baltimore to Chicago and return.....................	19.0	67,050	140	.21	30.0	29.0
	Car 2–B, held on track in Baltimore yards.......................		67,070	160	.24	30.0	31.7
3.....	Car 3–C, transit, Baltimore to Chicago and return.....................	17.0	57,080	50	.09	31.0	29.0
	Car 3–B, held on track in Baltimore yards.......................		56,890	20	.04	30.0	32.2
4.....	Car 4–C, transit, Baltimore to Chicago and return.....................	13.3	67,080	90	.13	34.0	29.0
	Car 4–B, held on track in Baltimore yards.......................		67,100	110	.16	35.0	33.5

As shown in Table IV and figure 3, the car containing corn of 22 per cent moisture which was shipped from Baltimore to Chicago and return lost 0.27 of 1 per cent, while the car containing the same kind of corn but which was held on the track at Baltimore lost 0.34 of 1 per cent. The corn from lot No. 2, containing 19 per cent of moisture, showed a loss in weight of 0.21 of 1 per cent for the car en route from Baltimore to Chicago and return, as compared with a loss of 0.24 of 1 per cent for the duplicate car held on the track at Baltimore. The reverse is true, however, of the corn containing 17 per cent of moisture. In this lot the shrinkage was 0.09 of 1 per cent in the corn in transit and only 0.04 of 1 per cent in the car of corn that remained on the track at Baltimore. It should also be noted in this connection that this lot of corn was of exceptional quality, showing the highest percentage of sound corn at the beginning of the experiment. The corn with 13.3 per cent of moisture lost 0.13 of 1 per cent in transit and 0.16 of 1 per cent on the track at Baltimore. As in the first experiment, the losses in weight of the four lots of corn used varied according to the moisture content of the corn, with the exception of lot No. 4. The fact that this lot of corn, which had an average moisture content of 13.3 per cent. lost more in weight than the corn that contained 17 per cent of moisture was probably due to the exceptionally good quality of the corn in lot No. 3, or possibly to the admixture of a larger percentage of finely broken corn in lot No. 4, although no leakage, even of the finely broken corn, could be detected in any part of the cars.

As shown in Table IV and figure 3. the temperature of the corn at the time of loading it into the cars which were shipped from Baltimore to Chicago and return ranged from 30° to 34°, with an average of 31.5° F. for the corn in all of the cars. At the time of unloading the range in temperature was from 29° to 33°, with an average of 30° F., or an average decrease of 1.5 degrees. The average temperature of the corn at the time of loading it into the cars which were held on the track in Baltimore ranged from 30° to 35°, with an average of 31.5° F. for the corn in all the cars. At the time of unloading the temperature ranged from 31.7° to 38.2°, with an average of 33.9° F. This is an average increase of 2.4 degrees, as compared with

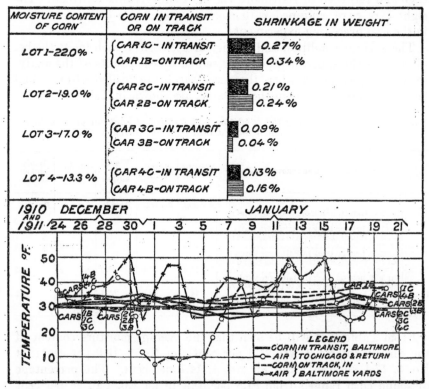

Fig. 3.—Diagram showing the loss in the weight of the corn, the average temperature of the corn in each of four cars in transit from Baltimore to Chicago and return, and also the mean daily air temperature through which the cars passed, as compared with the same factors on the four duplicate cars of corn held on the track at Baltimore in the second experiment.

an average decrease of 1.5 degrees for the cars moved from Baltimore to Chicago and return. This variation in the temperature of the corn was due to the wide difference in air temperature between Baltimore and the western points, as shown in figure 3, page 9.

The condition of the corn in the second experiment was practically the same at the end as at the beginning of the test. This is shown in figure 3 by the fact that the average temperature of the corn in the cars did not increase to any great extent from the time they were loaded until they were unloaded.

THIRD EXPERIMENT, MARCH 2 TO MARCH 29, 1911.

CORN LOADED INTO CARS.

On March 2, 1911, four lots of corn were selected and two cars were loaded from each of the four lots. This corn was selected and handled in the same manner as that in the second and fourth experiments, one car from each lot of corn being held on the track at Baltimore, while the duplicate car from each lot was forwarded to Chicago and returned to Baltimore via the Pennsylvania Railroad. The eight cars were equipped with electrical resistance thermometers as in the two preceding experiments.

The average condition and quality of the corn in each lot at the beginning of the experiment, as represented by the results of the analyses of samples taken when the corn was being loaded after it had been thoroughly mixed in the elevator, are shown in Table V.

TABLE V.—*Factors showing the condition and quality of each lot of corn in the third experiment when loaded into cars on March 2, 1911.*

Lot No.	Car designation and movement.	Moisture content.	Sound corn.	Weight per bushel.	Cob, dirt, etc.	Badly broken kernels.
		Per cent.	*Per cent.*	*Pounds.*	*Per cent.*	*Per cent.*
1.....	Car 1–C, transit, Baltimore to Chicago and return.	21.6	94.8	51.9	0.2	4.0
	Car 1–B, held on track in Baltimore yards...					
2.....	Car 2–C, transit, Baltimore to Chicago and return.	19.9	95.1	53.3	.2	3.6
	Car 2–B, held on track in Baltimore yards....					
3.....	Car 3–C, transit, Baltimore to Chicago and return.	17.4	97.0	55.9	.2	4.2
	Car 3–B, held on track in Baltimore yards....					
4.....	Car 4–C, transit, Baltimore to Chicago and return.	14.1	98.0	56.5	1.0	17.2
	Car 4–B, held on track in Baltimore yards....					

From Table V it will be seen that the average moisture content of the corn in lot No. 1 was 21.6 per cent; lot No. 2, 19.9 per cent; lot No. 3, 17.4 per cent; and lot No. 4, 14.1 per cent. The percentage of sound corn ranged from 94.8 per cent in lot No. 1 to 98 per cent in lot No. 4.

SHRINKAGE IN THE WEIGHT OF THE CORN.

The quantity of corn placed in each car, the shrinkage in weight while in the cars from March 2 to March 29, 1911, and the average temperature of the corn in each car at the time of loading and unloading are given in Table VI. These same factors, together with the daily mean temperature of the air through which the cars passed in transit from Baltimore to Chicago and return, and the daily average temperature of the corn in each of these cars, as compared with like records made in connection with the cars held on the track in the Pennsylvania Railroad yards, at Baltimore, are shown in figure 4.

TABLE VI.—*Weight of corn in each car immediately after loading on March 2, 1911, and the shrinkage, or loss in weight, together with the moisture content at time of loading and the average temperature of the corn at time of loading and unloading.*

Lot No.	Car designation and movement.	Moisture content of corn when loaded.	Net corn in car when loaded.	Shrinkage or loss in weight.		Average temperature of corn when—	
				Pounds.	Per cent.	Loaded.	Unloaded.
		Per cent.	*Pounds.*			° F.	° F.
1.....	Car 1–C, transit, Baltimore to Chicago and return	21.6	67,170	390	0.58	40.8	112.0
	Car 1–B, held on track in Baltimore yards		67,200	400	.59	40.0	109.7
2.....	Car 2–C, transit, Baltimore to Chicago and return	19.9	66,970	160	.24	40.0	41.5
	Car 2–B, held on track in Baltimore yards		66,740	175	.26	40.0	41.5
3.....	Car 3–C, transit, Baltimore to Chicago and return	17.4	56,600	130	.23	40.0	40.7
	Car 3–B, held on track in Baltimore yards		56,350	130	.23	40.0	40.5
4.....	Car 4–C, transit, Baltimore to Chicago and return	14.1	66,580	100	.15	40.0	41.3
	Car 4–B, held on track in Baltimore yards		66,250	70	.11	40.0	41.3

As shown in Table VI and figure 4, the shrinkage, or loss in weight, of the corn varied according to the moisture content. The corn containing 21.6 per cent of moisture lost 0.58 of 1 per cent when en route from Baltimore to Chicago and return, while the duplicate car of corn held on the track at Baltimore showed a shrinkage in weight of 0.59 of 1 per cent. The car of corn which contained 19.9 per cent of moisture lost 0.24 of 1 per cent in transit, while the duplicate lot in the car which was held on the track at Baltimore showed a loss of 0.26 of 1 per cent. The corn from lot No. 3, containing 17.4 per cent of moisture, showed the same shrinkage in transit as the car of corn which remained on the track at Baltimore—0.23 of 1 per cent. The corn from lot No. 4, containing 14.1 per cent of moisture, showed 0.15 of 1 per cent loss in weight while on the car in transit, as against 0.11 of 1 per cent loss on the corn in the duplicate car at Baltimore.

In this experiment the losses in weight were practically the same on the cars of corn in transit as on the cars held at Baltimore. By consulting the temperature records shown in figure 4 it will be seen that there was not such a marked difference in the air temperatures at Baltimore and at western points during the third experiment as occurred in the second experiment, which is illustrated in figure 3, page 9. The air temperature through which the cars passed en route from Baltimore to Chicago and return was lower on several days than the air temperature at Baltimore, but this difference occurred only for three or four days at most, and on several days the air temperature surrounding the cars in transit was higher than the air temperature at Baltimore. The difference between the air temperature at Baltimore and at western points during this experiment was so slight that it did not affect the temperature of the corn appreciably, and consequently the losses in weight were practically the same on the corn in transit as on the corn at Baltimore.

The corn in the two cars loaded from lot No. 1 of the third experiment showed a marked deterioration at the end of the test. This

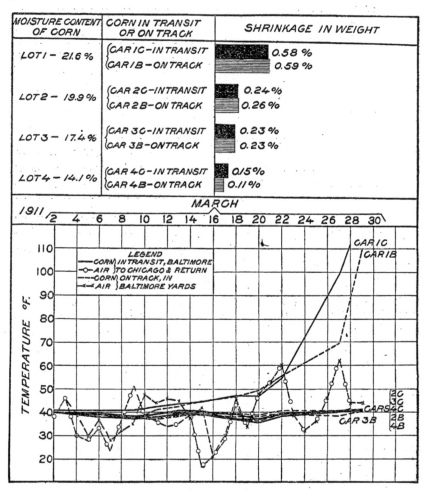

Fig. 4.—Diagram showing the loss in the weight of the corn, the average temperature of the corn in each of four cars in transit from Baltimore to Chicago and return, and also the mean daily air temperature through which the cars passed, as compared with the same factors on the four duplicate cars of corn held on the track at Baltimore in the third experiment.

deterioration was due almost entirely to the high moisture content of the corn, which became hot in the cars as a result of fermentation and the development of molds. On March 22 the average temperature of the corn in each of the two cars was approximately 55° F. At the end of the experiment the average temperature of the corn in the car that remained on the track at Baltimore had increased to

109.7°, while that of the corn in transit had increased to 112° F. The high temperature accompanying the deterioration accounts for the very large natural shrinkage occurring in these cars.

The corn in the other six cars remained at practically the same temperature throughout the experiment, showing only a slight increase from the time of loading until the time of unloading. This slight increase in temperature was probably due to the fact that the atmosphere was generally warmer during the latter part of the experiment rather than to any increase in temperature resulting from fermentation, as the corn was in good condition when unloaded.

FOURTH EXPERIMENT, MAY 11 TO JUNE 1–3, 1911.

CORN LOADED INTO CARS.

On May 11, 1911, eight cars were loaded with corn, two cars each from four special lots which had been selected according to moisture content. Prior to loading into the cars, each lot was thoroughly mixed in the elevator. One car of corn from each lot was held on the track at Baltimore, while the duplicate car of each lot was forwarded to Chicago and returned to Baltimore via the Pennsylvania Railroad. The cars were equipped with electrical resistance thermometers, which were located in the grain in the same manner as in the three former experiments.

The average condition and quality of the corn in each lot at the beginning of the experiment, as represented by the results of the analyses of samples taken when the corn was being loaded, are shown in Table VII.

TABLE VII.—*Factors showing the condition and quality of each lot of corn in the fourth experiment when loaded into cars on May 11, 1911.*

Lot No.	Car designation and movement.	Moisture content.	Sound corn.	Weight per bushel.	Cob, dirt, etc.	Badly broken kernels.
		Per cent.	Per cent.	Pounds.	Per cent.	Per cent.
1	Car 1–C, transit, Baltimore to Chicago and return... Car 1–B, held on track in Baltimore yards....	18.2	89.9	54.0	0.2	3.0
2	Car 2–C, transit, Baltimore to Chicago and return... Car 2–B, held on track in Baltimore yards....	17.8	91.1	54.1	.2	3.2
3	Car 3–C, transit, Baltimore to Chicago and return... Car 3–B, held on track in Baltimore yards....	16.9	95.7	55.5	.3	4.4
4	Car 4–C, transit, Baltimore to Chicago and return... Car 4–B, held on track in Baltimore yards....	13.9	96.2	56.2	.7	22.1

It will be seen by referring to Table VII that the average moisture content of the corn in lot No. 1 was 18.2 per cent. The corn in this lot was considerably lower in moisture than the corn in lot No. 1 of the second and third experiments. This was made necessary because it was almost impossible to obtain corn having a moisture content of more than 18 or 19 per cent that was sound and sweet at that

season of the year. The average moisture content of the corn in lot
No. 2 was 17.8 per cent. In lot No. 3 the average moisture content
was 16.9 per cent, while in lot No. 4 the average moisture content
was 13.9 per cent. The average proportion of sound corn in each lot
ranged from 88.9 per cent in lot No. 1 to 96.2 per cent in lot No. 4.
It will be noted that the average percentage of badly broken corn was
very high in lot No. 4, being 22.1 per cent.

SHRINKAGE IN THE WEIGHT OF THE CORN.

The quantity of corn placed in each car, the shrinkage in weight
while the corn was contained in the cars, from May 11 to June 1 and

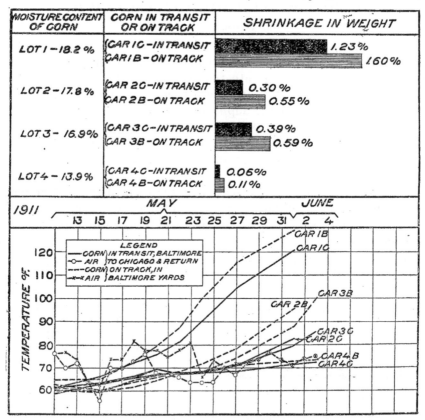

FIG. 5.—Diagram showing the loss in the weight of the corn, the average temperature
of the corn in each of four cars in transit from Baltimore to Chicago and return, and
also the mean daily air temperature through which the cars passed, as compared with
the same factors on the four duplicate cars of corn held on the track at Baltimore in
the fourth experiment.

3, 1911, and the average temperature of the corn in each car at the
time of loading and unloading are given in Table VIII. These same
factors, together with the mean daily air temperatures to which the cars
were exposed and the average temperature of the corn in the various
cars throughout the experiment, are shown more in detail in figure 5.

TABLE VIII.—*Weight of corn in each car immediately after loading on May 11, 1911, and the shrinkage, or loss in weight, together with the moisture content at time of loading and the average temperature of corn at time of loading and unloading.*

Lot No.	Car designation and movement.	Moisture content of corn when loaded.	Net corn in car when loaded.	Shrinkage, or loss in weight.		Average temperature of corn when—	
				Pounds.	Per cent.	Loaded.	Unloaded.
		Per cent.	*Pounds.*			° *F.*	° *F.*
1.....	Car 1–C, transit, Baltimore to Chicago and return................	18. 2	69,850	850	1. 23	61. 7	121. 2
	Car 1–B, held on track in Baltimore yards..................		69,470	1,110	1. 60	64. 2	129. 0
2.....	Car 2–C, transit, Baltimore to Chicago and return................	17. 8	69,910	210	. 30	58. 7	82. 8
	Car 2–B, held on track in Baltimore yards..................		68,750	370	. 55	60. 0	95. 7
3.....	Car 3–C, transit, Baltimore to Chicago and return................	16. 9	56,060	220	. 39	61. 0	85. 7
	Car 3–B, held on track in Baltimore yards..................		55,700	320	. 59	61. 8	100. 2
4.....	Car 4–C, transit, Baltimore to Chicago and return................	13. 9	65,890	40	. 06	61. 3	72. 5
	Car 4–B, held on track in Baltimore yards..................		63,430	70	. 11	62. 5	.73. 3

From Table VIII and figure 5 it will be seen that the shrinkage in weight was much greater in the cars of corn that were held on the track at Baltimore than in those in transit. The mean daily air temperature through which the cars passed while en route, as shown in figure 5, was considerably less than the mean daily air temperature that surrounded the cars at Baltimore. This caused the temperature of the corn in the cars on the track at Baltimore to increase faster than that of the corn in transit, in that the higher air temperature not only warmed the corn, but also offered much more favorable conditions for the development of molds and bacteria and for the action of enzyms or other unorganized ferments; consequently, the natural shrinkage in the weight of the corn held at Baltimore was greater than that of the corn in transit.

The shrinkage in weight of the car of corn in transit representing lot No. 1, containing 18.2 per cent of moisture, was 1.23 per cent, while the duplicate car that remained on the track at Baltimore lost 1.6 per cent. The heavy losses occurring in these two cars were directly due to the deterioration of the corn. As may be seen in figure 5, on May 20 the average temperature of the corn from lot No. 1 was 79.5° in the car held on the track at Baltimore and 75.3° in the car in transit. From May 20 until June 1 the temperature of the corn increased rapidly, and on the latter date the car of corn held at Baltimore showed a temperature of 129°, while the corn that had been shipped from Baltimore to Chicago and return showed an average temperature of 121.2°. The corn in these two cars was in a very badly damaged condition when unloaded, on June 1.

In the second lot the loss in the weight of the corn shipped to Chicago and returned to Baltimore was 0.3 of 1 per cent. There was no appreciable increase in the temperature of this corn until May 25, at which time the average temperature of the corn in the car on the track at Baltimore was 71.3°. On June 1 the average temperature had increased to 95.7°. The corn that was in transit to Chicago showed an average of only 82.8° at the time of unloading, on June 1. This corn was only slightly damaged. It had, however, reached a point where the deterioration and the accompanying shrinkage would have been very rapid within the next few days. The car of corn from lot No. 3, containing an average moisture content of 16.9 per cent, lost 0.39 of 1 per cent in transit, while the loss on the same kind of corn in the duplicate car held on the track at Baltimore was 0.59 of 1 per cent. This corn was not unloaded until June 3, two days after the corn from the first and second lots was unloaded. At the time of unloading, the temperature of the corn in the car which was returned from Chicago was 85.7°, while the temperature of the corn in the car that remained on the track at Baltimore was 100.2°. As a result of the delay of two days in the time of unloading, the shrinkage in lot No. 3, containing 16.9 per cent of moisture, was greater than the shrinkage in lot No. 2, which contained 17.8 per cent of moisture. If the corn from lot No. 3 had been unloaded on June 1, the loss in weight of the corn that was shipped to Chicago would have been 0.29 of 1 per cent and the duplicate car at Baltimore would have shown a loss of 0.44 of 1 per cent. The loss in weight for those two days was 60 pounds in the car of corn returned from Chicago and 80 pounds for the car on the track. The average temperature of the two cars of corn from lot No. 3 on June 1 was also lower than the temperature of the two cars of corn from lot No. 2. In this connection it is important to note that the corn in lot No. 3 was fresh-shelled corn, while that in lot No. 2 was shelled in November or December, 1910. This factor had a considerable influence on the degree of deterioration and consequently had an effect on the amount of shrinkage. The car of corn from lot No. 4, containing 13.9 per cent of moisture, which was shipped from Baltimore to Chicago and return, lost 0.06 of 1 per cent. The duplicate car from the same lot held on the track at Baltimore lost 0.11 of 1 per cent.

SHRINKAGE AS AFFECTED BY THE TEMPERATURE OF THE CORN.

In all cases the temperature of the corn immediately after loading into the cars was practically the same as the air temperature at the time of loading.

Table IX gives the total losses in weight of the corn in all the cars in each experiment in transit, as compared with the corn from

the same lots on the track at Baltimore; also the average temperature of the corn when loaded and just prior to unloading, together

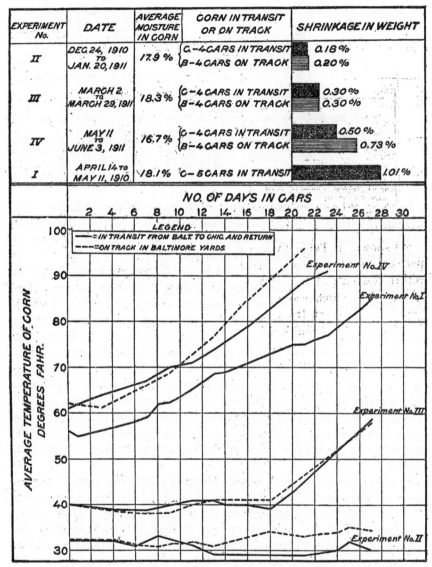

EXPERIMENT No.	DATE	AVERAGE MOISTURE IN CORN	CORN IN TRANSIT OR ON TRACK		SHRINKAGE IN WEIGHT
II	DEC. 24, 1910 TO JAN. 20, 1911	17.9 %	C-4 CARS IN TRANSIT		0.18 %
			B-4 CARS ON TRACK		0.20 %
III	MARCH 2 TO MARCH 29, 1911	18.3 %	C-4 CARS IN TRANSIT		0.30 %
			B-4 CARS ON TRACK		0.30 %
IV	MAY 11 TO JUNE 3, 1911	16.7 %	C-4 CARS IN TRANSIT		0.50 %
			B-4 CARS ON TRACK		0.73 %
I	APRIL 14 TO MAY 11, 1910	18.1 %	C-5 CARS IN TRANSIT		1.01 %

FIG. 6.—Diagram showing the average loss in weight and the average temperature of
the four cars of corn in transit from Baltimore to Chicago and return in each of the
four experiments, compared with the average loss in weight of the four cars of corn
from the same lots that were held on the track at Baltimore and the average moisture
content of each of the four lots at the beginning of the experiments.

with the average moisture content. These same factors are also
shown diagrammatically in figure 6.

TABLE IX.—*Total amount of corn contained in each experiment, the average moisture content of the corn at the beginning of the experiment, and the average temperature of the corn immediately after loading and just prior to unloading.*

Experiment No.	Period covered by experiment.	In transit from Baltimore to Chicago and return.	Held on track in Baltimore yards.	Moisture content.	Net corn in cars after loading.	Shrinkage or loss in weight.		Average temperature of corn when—	
						Pounds.	Per cent.	Loaded.	Unloaded.
				Per cent.	*Pounds.*			°*F.*	°*F.*
1.......	Apr. 14 to May 11, 1910.	5 cars.	....	18.1	334,270	3,360	1.01	55.0	85.2
2.......	Dec. 24, 1910, to Jan. 20, 1911.	4 cars.	....	17.9	258,350	460	.18	31.5	30.0
		....	4 cars.	17.9	258,200	520	.20	31.5	33.9
3.......	Mar. 2 to Mar. 29, 1911.	4 cars.	....	18.3	257,320	780	.30	40.2	58.9
		....	4 cars.	18.3	256,540	775	.30	40.0	58.3
4.......	May 11 to June 1–3, 1911.	4 cars.	....	16.7	261,710	1,320	.50	60.7	90.6
		....	4 cars.	16.7	257,350	1,870	.73	61.6	99.6

In all of the experiments, with the exception of the first, four cars of corn differing in moisture content were shipped from Baltimore to Chicago and return and four duplicate cars of the same kind of corn were held on the track at Baltimore.

It will be seen in Table IX and figure 6 that the average loss in weight of the four cars of corn in transit from Baltimore to Chicago and return in the second experiment, begun December 24, 1910, was 0.18 of 1 per cent. The four duplicate cars held at Baltimore showed an average shrinkage of 0.20 of 1 per cent. In figure 6 it will be noticed that the average temperature of the corn in the eight cars at the time of loading was 31.5° F. At the time of unloading there was a difference in temperature between the corn in transit and that on the track of 3.9 degrees, the temperature of the corn in transit having decreased 1.5 degrees, while that of the corn on the track had increased 2.4 degrees. This was caused by the lower air temperatures in transit than at Baltimore, as shown in figure 3, page 9.

In the third experiment, begun March 2, 1911, the losses were the same on the corn in transit as on the corn at Baltimore, being 0.3 of 1 per cent. In this experiment the average temperature of all the corn was 40° F., or 8½ degrees higher than the corn used in the second experiment. The average increase in the temperature of the corn in transit from the time of loading until it was unloaded was 18.7 degrees. The average increase in the temperature of the corn on the track was 18.3 degrees. By consulting figure 4 it will be seen that the air temperature surrounding the corn in transit was practically the same as the air temperature surrounding the corn at Baltimore. In both cases the tendency was toward a higher air temperature; consequently, the temperature of the corn in transit increased uniformly with the temperature of the corn at Baltimore.

In the fourth experiment the temperature of all the corn at the time of loading, May 11, 1911, was 21.2 degrees higher than the temperature of the corn in the third experiment, being 61.2° and ranging from 60.7° in the corn in transit to 61.6° F. in the corn on the track. From figure 5 it will be seen that during the fourth experiment the air temperature at western points was always lower than the air temperature at Baltimore; thus, the corn in the four cars on the track showed an average temperature of 99.6° at the end of the experiment, as against an average of 90.6° F. for the corn in the four cars in transit. The average loss in weight of the corn in transit was 0.50 of 1 per cent, and of the corn held on the track 0.73 of 1 per cent.

In the first experiment, conducted from April 14 to May 11, 1910, no duplicate cars were held on the track at Baltimore. The average loss in weight of the corn in the five cars in this experiment was 1.01 per cent. The average temperature of the corn at the time of loading was 55° F., or 6.2 degrees lower than the average temperature of the corn used in the fourth experiment, May 11, 1911. The fact that the corn in the first experiment showed a greater shrinkage than the corn in the experiment conducted from May 11 to June 1–3, 1911, when the average temperature of the corn and the average of the mean daily air temperatures were considerably higher, is explained by the fact that the corn in the first experiment showed an average moisture content of 18.1 per cent as against 16.7 per cent for the corn in the fourth experiment. The corn was also stored in the cars for five days longer; therefore, the difference in moisture and the difference in the number of days that the corn was contained in the cars caused a larger percentage of shrinkage in the corn that was in transit from April 14 to May 11, 1910, than was found in the experiment conducted from May 11 to June 1–3, 1911.

Figure 6 also shows that the average moisture content of the corn in the shipment made on December 24, 1910, was 17.9 per cent. The number of days the corn was contained in the cars was 27. The shipment made March 2, 1911, contained corn with an average moisture content of 18.3 per cent. This corn was also stored in the cars for 27 days. The average moisture content of the corn in the shipment of April 14, 1910, was 18.1 per cent, the duration of this test being likewise 27 days. However, as corn with a higher moisture content deteriorates very rapidly during the spring months, when the weather is generally warm, corn with a moisture content of 16.7 per cent was selected for the shipment of May 11, 1911, and the time was reduced from 27 to 22 days. Even in this experiment, as shown in figure 5, the corn in two of the cars was hot and very badly damaged, while the corn in four of the remaining six cars had begun to heat before unloading.

COMPARISONS OF THE SAME LOT OF CORN USED IN THE SECOND, THIRD, AND FOURTH EXPERIMENTS.

The lot of corn containing 17 per cent of moisture used in the experiment begun on December 24, 1910, was also used in the experiments begun on March 2 and on May 11, 1911. One car of this lot of corn remained on the track at Baltimore and one car of the same lot was shipped to Chicago and returned to Baltimore in each of the three experiments.

In figure 7 a comparison is made of the loss in weight of the corn in this lot in each experiment. In the experiment begun on December 24, 1910, when the temperature of the corn was below freezing, ranging from 30.5° at the time of loading to 30.6° F. at the time of

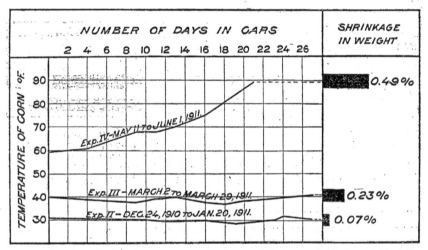

Fig. 7.—Diagram showing the average temperature and the average loss in weight of the one lot of corn used in the three experiments begun on December 24, 1910, and March 2 and May 11, 1911.

unloading, the shrinkage was 0.07 of 1 per cent. This same corn in the shipment made March 2, 1911, showed a temperature at the time of loading of 40° and at the time of unloading of 40.6° F. In this case the shrinkage was 0.23 of 1 per cent. In the shipment made on May 11, 1911, the average temperature of the corn at the time of loading was 61.4°, while at the time of unloading the average temperature was 93° F. The average percentage of shrinkage was 0.49 of 1 per cent.

SUMMARY OF THE SECOND, THIRD, AND FOURTH EXPERIMENTS.

A summary of the results of the second, third, and fourth experimental shipments, with special reference to the average moisture content, the percentage of shrinkage, and the temperature of the corn at the time of unloading the 12 cars in transit from Baltimore to

Chicago and return, in comparison with similar data covering the duplicate lot of 12 cars that were held on the track at Baltimore, is shown in figure 8. From this figure is will also be seen that the average natural shrinkage in the 12 cars of corn in transit amounted to 0.33 of 1 per cent, while the average natural shrinkage in the 12 duplicate cars of corn that were held on the track at Baltimore was 0.41 of 1 per cent. The average temperature at the time of unloading the corn in the 12 cars that were shipped to Chicago and returned to Baltimore was 60°, as against 64° F. for the 12 duplicate cars held on the track at Baltimore. This difference of 4 degrees is accounted for by the fact that the temperature of the air through which the cars passed en route to Chicago and return was usually lower than the temperature of the air surrounding the cars at Baltimore. The

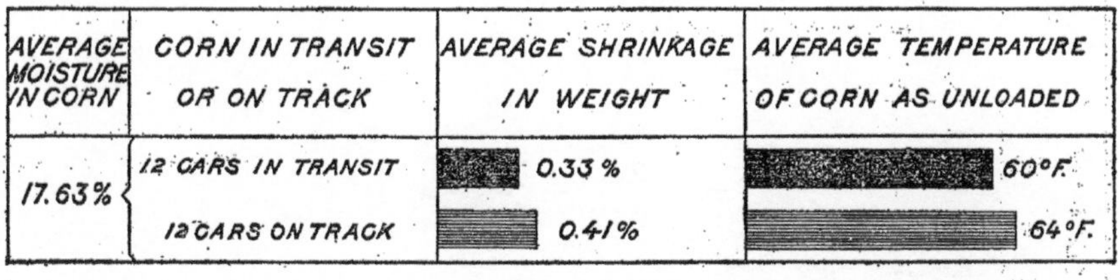

AVERAGE MOISTURE IN CORN	CORN IN TRANSIT OR ON TRACK	AVERAGE SHRINKAGE IN WEIGHT	AVERAGE TEMPERATURE OF CORN AS UNLOADED
17.63% {	12 CARS IN TRANSIT	0.33 %	60°F.
	12 CARS ON TRACK	0.41%	64°F.

FIG. 8.—Diagram showing the average loss in weight of the corn, the average moisture content of the corn at the time of loading, and the average temperature of the corn at the time of unloading in the 12 cars that were shipped to Chicago and returned to Baltimore in the three experiments begun on December 24, 1910, and March 2 and May 11, 1911, as compared with the same factors on the 12 duplicate cars of corn held on the track at Baltimore.

average of the mean daily air temperatures through which the cars passed en route to Chicago and return was 46.2°, and the average of the mean daily air temperatures surrounding the cars at Baltimore was 50.9° F., or 4.7 degrees higher at Baltimore than at western points.

CONCLUSIONS.

(1) There is unquestionably a natural shrinkage in commercial corn during transit and while in storage.

(2) Natural shrinkage varies with the moisture content of the corn and the atmospheric conditions to which it is exposed.

(3) Natural shrinkage in corn that has become sour and hot is very rapid and may amount to several per cent within a few days.

O